Managing Our Resources

Plants

A resource our world depends on

Heinemann Library
Chicago, Illinois

Ian Graham

a division of Reed Elsevier Inc.

Chicago, Illinois

Customer Service 888-454-2279

Visit our website at
www.heinemannlibrary.com

Designed by David Poole and
Paul Myerscough
Photo research by Melissa Allison and
Andrea Sadler

Originated by Ambassador Litho Ltd.
Printed in China by WKT Company Limited

09 08 07 06 05

10 9 8 7 6 5 4 3 2 1

Library of Congress Cataloging-in-Publication Data
Graham, Ian, 1953-
Plants : a resource our world depends on / Ian Graham.
p. cm. -- (Managing our resources)
Includes bibliographical references and index.
ISBN 1-4034-5619-4 (hc : lib. bdg) --
ISBN 1-4034-5627-5 (pb)
1. Plants--Juvenile literature. I. Title. II. Series.
QK49.G64 2004
333.95'3--dc22

2004005905

Acknowledgments
The author and publisher are grateful to the following for permission to reproduce copyright material: p. 4 W. Broadhurst/FLPA; p. 5 top Papilio Neil Miller/Ecoscene; p. 5 bottom Corbis; p. 6 Jurgen & Christine Johns/FLPA; p. 7 top Jeremy Walker/Science Photo Library; p. 7 bottom Tony Page/Ecoscene; p. 8 Sidney Moulds/Science Photo Library; p. 9 top Anthony Cooper/Ecoscene;p. 9 bottom Photodisc/Getty Images; p. 10 Hans Dieter Brandl/FLPA; p. 11 Catherine Mullen/FLPA; p. 12 Joel Creed/Ecoscene; p. 13 top D. Hall/FLPA; p. 13 bottom Mary Evans Picture Library ; p. 14 Alex Barter/Science Photo Library; p. 15 Herman Eisenbess/Science Photo Library; p. 16 J. Watkins/FLPA; p. 17 Topham Picturepoint; p. 18 Geoff Tompkinson/Science Photo Library; p. 19 Stephen Dalton/NHPA; p. 20 top Simon Fraser/Science Photo Library; p. 20 bottom D. Warren/FLPA; p. 21 Tom Bean/Corbis; pp. 22 inset, 23 Science Photo Library; p. 24 Alan Towse/Ecoscene; p. 25 Andrew Brown/Ecoscene; p. 26 Maurice Nimmo/FLPA; p. 27 top Simon Grove/Ecoscene; p. 27 bottom Michael Pole/Corbis; p. 28 Andy Hibbert/Ecoscene; p. 29 Royal Botanical Gardens, Kew.

Cover photograph: Corbis/First Light

Contents

Some words are shown in bold, **like this.** You can find out what they mean by looking in the glossary.

What Are Plants?

Plants are living things that grow in gardens and in nature. Some plants coat the ground with a green carpet, while others stand up above the ground. Plants can be so small that you cannot see them, or they can be trees as tall as buildings. Stems or trunks hold leaves up in the sunshine. Under the ground, their roots grow down into the soil and hold them firmly in place.

Why are most plants green?

The green parts of trees and other plants use sunlight to change simple materials from the soil and the air into food. This **process** is called **photosynthesis.** The substance in plants that makes photosynthesis possible is called chlorophyll. Sunlight contains all the colors of the rainbow. Chlorophyll soaks up some colors but reflects others, mainly green. That is why most plants are green. Red plants have chlorophyll, too, but it is masked by other colors.

Plants of all shapes and sizes cover the ground.

Did you know?

Some plants get extra food by catching insects. The plant juices dissolve, or break down, the trapped insects, and the plant takes in the nutritious liquid.

The Venus's flytrap lives on insects that it catches by snapping its pair of leaves together.

Fungi

Some things we think of as plants are actually **fungi.** They live on dead and rotting plants, or they survive by living in food made by living plants. Because fungi do not use sunlight to make food, they have no chlorophyll and so they are not green. Fungi include mold, mushrooms, and toadstools.

Why Are Plants Important?

Plants are vital for life. When green plants use **photosynthesis** to make food, they take in carbon dioxide gas from the air and give out oxygen gas. The oxygen becomes part of the air around us. People and other animals breathe the air in, take in the oxygen, and breathe out carbon dioxide.

Why do we eat plants?

Animals, including humans, cannot make food for themselves in the way that plants do, so we let plants make food and then we eat the plants. Luckily for us, plants make more food than they need. They store the extra food in their roots, stems, seeds, or fruit. Potatoes, carrots, peas, and apples are all examples of food stored in plants that we can eat.

Did you know?

A lot of plants are poisonous and must not be eaten. If you do not know what a plant is, do not eat it!

Most animals depend on plants for food. They either eat plants or they eat animals that live on plants.

What Are Plants Used For?

Plants have a great variety of uses for humans other than giving us food. They contain fibers that can be woven into textiles, which is material for clothing. For example, cotton comes from the cotton plant, and linen comes from the flax plant. Canvas is made from cotton or jute plants. Sisal, from the agave plant, and jute fibers are used to make ropes. Some medicines are made from substances **extracted** from plants. Timber, from tree trunks, is an essential building material. It is used for building because it is strong and also easy to cut and shape. Plants can even supply fuel for engines.

One of the most widespread and important uses of plants is for food, like the wheat in this photograph.

Bamboo

One very useful plant is a type of grass called bamboo. Its seeds and **shoots** can be eaten. Paper can be made from it. Large bamboo plants have thick, pipelike stems that are strong enough to be used for building.

Forests of fast-growing trees are specially grown to supply timber for the construction of houses and buildings.

How are plants used in medicine?

People have known for thousands of years that some plants could treat health problems. Treating illnesses with plants is called herbal medicine. Herbal medicine is still widely used today. Some modern medicines also use plants, but the medicines are produced by scientific methods. Aspirin was one of the first modern medicines, and it is still commonly used today. The pain-killing chemical in aspirin comes from willow trees. Penicillin, a medicine that kills some **organisms** that cause disease, comes from mold, which is a type of **fungus.**

Some medicines are made from molds that grow on stale food.

How do plants keep us healthy?

A good diet contains everything we need to stay healthy. Small amounts of vitamins and minerals are essential. Some of these come from meat and dairy products, but many of them come from fruits, vegetables, and **cereals.** Plants also give us fiber, or roughage. This helps the body's muscles move food through the body.

What are flowers used for?

We use flowers in many of the ceremonies that mark important life events, such as weddings and funerals. Vases of flowers brighten people's homes and offices. Growing flowers to meet this demand is a huge business in the United States, the Netherlands, and Japan. Cold storage and fast air transportation enable growers to send fresh flowers all over the world within a day or two.

Flower markets supply a great variety of flowers from all over the world.

Did you know?

Saffron is one of the most expensive spices used in cooking. It is made from part of the crocus flower. About 170,000 flowers have to be picked by hand to produce just 2 pounds (1 kilogram) of saffron!

Why are plants good for the land?

As trees and other plants grow, their roots spread through the ground and help to stop the soil from being blown or washed away. Road and railway embankments are planted with grass to hold the soil together.

Coastal erosion

Coastal erosion occurs when the wind and the sea wear away part of the coast. It can be reduced by planting grass along the seashore. Most grass cannot live in sand where saltwater is present. However, a type of grass called marram grass, or beach grass, can live in sand near the sea. This tough grass has underground stems that can grow up to 50 feet (15 meters) long and can send up **shoots** at intervals. The network of stems along with the grass's roots help to hold sand together.

Marram grass holds sand dunes together.

What is biomass?

Biomass fuel is made from plant and animal waste. Trees and other green plants take in energy from the Sun all through their life spans. We use this energy by burning the plants. The energy stored inside them changes into heat and light.

Biomass fuel was the only fuel that people could burn to cook food and warm themselves for thousands of years. It is still the main fuel used in most parts of the developing world. Today, **fossil fuels** such as coal, oil, and natural gas are used, but these fuels will not last forever. Biomass fuel is a renewable energy source—it can be regrown and so it will not run out. As a result, biomass fuel is now being burned in some power plants to generate electricity.

Wood has been used as a fuel for fires for thousands of years.

How can plants power cars?

Plants contain starch and sugar, which can be **processed** to make other substances. One of these substances, ethanol, can be burned in an engine instead of gas. It can also be added to other fuels to make them last longer. A mixture of gas and ethanol is called gasohol. Countries with little or no oil of their own are interested in using ethanol to reduce their energy costs. They can grow plants to make fuel instead of having to buy it from other countries.

This gasohol station is located in Brazil.

Scientists recently discovered a way to produce hydrogen gas by using a substance that comes from plants. Hydrogen can be used as a fuel. Today, hydrogen is made from natural gas. However, in the future, forests and crops such as sugar beets might be grown to make hydrogen fuel for cars.

How are trees used to make balloons?

Party balloons are made from a natural rubber called latex, which comes from trees. When the bark of a rubber tree is cut, milky **sap** called latex oozes out. It is collected and used to make many things, including party balloons.

Party balloons and surgical gloves are made with latex from rubber trees.

How are the colors in plants used?

The stems, leaves, roots, and fruits of some trees can be used to make natural dyes to color cloth. Nettle leaves produce a green dye. Dark blue berries from the elder plant produce a blue dye. A black dye can be made from the bark of the alder tree.

Did you know?

In ancient Britain, warriors painted their bodies with a blue dye called woad before they went into battle. The dye was made from the woad plant.

Where Are Plants Found?

Plants are found nearly everywhere on Earth. Most plants grow best in certain conditions. Cacti grow best in dry conditions. Seaweed and other aquatic plants grow underwater. Pine trees are best suited to high, cold places. Bog plants grow well in waterlogged soil. There are plants for almost every type of habitat, or natural home for a plant or animal, on Earth.

A place where many similar plants live in the same type of conditions is called a biome, or major life zone. A **rain forest** is a type of biome. Desert is another biome.

Cacti can live in ground that is too dry for most plants.

Did you know?

Trees live the longest of all plants. Most oak trees live up to 300 years, but some have lived for 1,500 years. Giant sequoia trees can live 2,500 years. The oldest living plant on Earth is a bristlecone pine tree in California that is more than 4,765 years old.

Do all plants grow in soil?

Most plants grow in soil, but a lot of plants can grow in places where there is no soil. Lichens grow on rock. They take in water and nutrients from rainwater. Mistletoe grows on trees. Its roots grow into the tree and **extract** water and food. Mosses grow in damp places, even if there is no soil. Plants called bromeliads can grow high up in rain forests on the sides of trees.

Did you know?

Some plants help each other. Orchid plants have a **fungus** living on their roots. Orchids are not very good at producing food. The fungus does it for them. In return, the orchid provides a home for the fungus. This partnership between different living things is known as symbiosis.

How do plants spread?

Flowering plants, including trees, multiply by producing seeds, which grow into new plants. A fine, yellow dust called pollen is produced by the male parts of flowers. Tiny bits, or particles, of pollen are carried away by insects or the wind. Some of the particles land on other flowers and join with the flowers' female parts to form seeds. Many plants are also able to grow a whole new plant from parts of their own stems or roots.

Some seeds are small enough to blow away . Others are eaten by birds. Seeds often pass undamaged through birds and, later, drop to the ground. Some seeds stick to the coats of passing animals that carry them away. In these ways, new plants can grow far away from their parent plants.

Trees produce new trees by making seeds, such as these seeds on a sycamore tree.

How Are Plants Processed?

Plants are often used just as they exist in nature. They are also **processed** in many different ways. The fruits and vegetables they produce are peeled, boiled, roasted, and pressed to make a variety of foods and drinks. Wheat grain is ground to produce flour for making bread. Heating hard vegetables such as potatoes and carrots makes them softer and easier to eat. Fruits such as oranges and pineapples are pressed to squeeze out juice for making drinks. Tea is made from dried leaves, and coffee is made by roasting beans from coffee bushes.

Fibers are **extracted** from plants for making paper and textiles. Plants are also treated with chemicals to make special ingredients for medicines.

Cooking hard vegetables makes them softer and easier to eat.

CASE STUDY: Papermaking

Paper for newspapers, magazines, books, writing pads, envelopes, tissues, and toilet paper is made from trees. First, tree trunks are chopped up. Then, they are ground into a thick, oatmeal-like substance called pulp. The pulp may be bleached to stop the paper from turning yellow. A machine spreads the pulp onto a moving belt. The belt has many tiny holes in it. Water drains out through the holes. The mat of fibers made by the pulp passes between rollers that squeeze out more water. Then, other rollers dry it out and smooth it. The paper comes out of the machine and is wound onto giant rolls.

Did you know?

A papermaking machine can make more than 100 feet (30 meters) of paper a second.

There is a constant need for wood from trees to make paper.

How Can Plants Be Harmed?

Plants can be harmed by many things, including insects, animals, the weather, and some substances that come from cars and factories. Plants can also catch diseases that can harm or kill them. Plants can also be harmed by people collecting them from nature.

How do insects harm plants?

Insects and other creatures want the food that plants make. Some insects, such as locusts, can eat entire plants. Other insects bite into plants and suck out their watery **sap.** In nature, plants and their insect pests live in balance with each other. Even if insects do not kill a plant, the damage they do can let in **viruses, fungi,** and bacteria, which are **organisms** that can cause disease. Farmers try to reduce the damage pests cause by treating crops with materials called pesticides.

Aphids are plant pests that feed by sucking the watery sap out of plants.

How does acid rain harm plants?

Gases produced by engines and power plants join with moisture in the air to make a harmful substance called **acid.** This forms acid rain. When acid rain falls on plants, it damages the sensitive tips of their **shoots** and roots. It also damages their leaves and lets in **organisms** that cause disease. Acid rain can harm an entire forest.

These trees have been damaged by acid rain.

How does weather affect plants?

Trees and other plants can be harmed by bad weather. Strong winds can blow down trees. A great storm that swept across England in 1987 uprooted 15 million trees! Frost can damage tender buds of new leaves and flowers that are beginning to grow. Heavy rain can flatten crops. Most plants depend on a constant supply of water, so **drought** can cause them to be damaged or destroyed.

Plants are easily damaged or destroyed by extreme weather.

CASE STUDY: North America's Prairies

North America once had huge grasslands called prairies. They covered large parts of the continent and stretched as far as the eye could see. They were created by **droughts,** grazing animals, and fires that stopped trees from growing and forming forests. The land, plants, animals, and **Native American** people lived together in a natural balance. Then, European settlers arrived. They turned the **fertile** prairie land into farmland and killed many of the grazing animals. Almost all of the prairies were destroyed.

Now, some prairies are being re-created. The **native** prairie plant **species** are being replanted and animals are grazing on the land again. One advantage of restoring prairie land is that, unlike farmland, it does not need chemicals to control weeds or insect pests.

Parts of North America were once covered by prairies like this.

What Are Genetically Modified Plants?

Trees and other plants are made of cells. Most plant cells are too small to see without a microscope. How the plant cells develop depends on a variety of factors. Each plant cell has **genes** that play a part in a plant's growth. The genes act like instructions that a plant follows as it grows. These instructions make one seed grow into a grass plant and another grow into an oak tree.

Scientists have learned how to identify the genes in a plant's cells. They have also learned how to change them. A plant with genes that have been changed by scientists is called a genetically modified **organism,** or GMO.

This is a microscopic view of cells in a leaf.

Why do scientists change plant genes?

Traditionally, plant **breeders** produced plants by collecting plant seeds and then growing the seeds. By breeding only from the biggest plants, the breeders can produce plants that are bigger. Similarly, by breeding only from plants that produce very large crops of fruit, they can grow new plants that produce even better crops. This is called selective breeding because the growers select which plants to breed from.

Some scientists try to understand which plant genes make a plant bigger, more fruitful, or resistant to diseases. If they know this, they can change the genes and produce exactly the plants that they want. Changing a plant's genes is called genetic engineering.

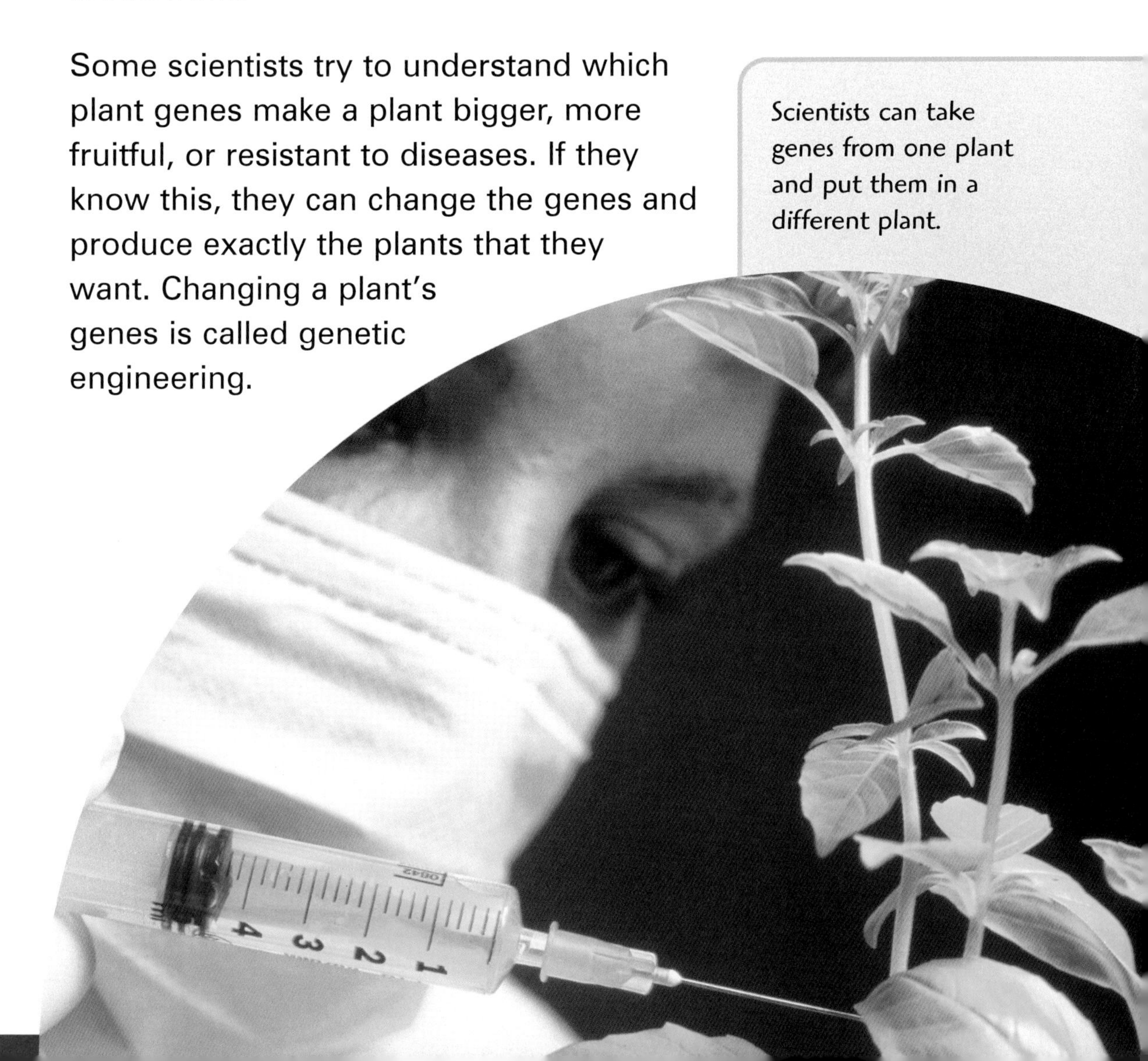

Scientists can take genes from one plant and put them in a different plant.

Is genetic engineering good or bad?

Some people think changing plants by using genetic engineering is a productive scientific practice. Others consider it to be unnatural.

In favor of genetic engineering

Scientists want to improve plants to feed more people. For example, scientists have produced a new type of rice that grows well in cold weather and in drier, salty soil. This rice can feed people in places where rice normally does not grow well. Other plants are being improved in the same way.

Against genetic engineering

Some people fear that new **genes** being put into crops will spread into wild plants and change them, too. They fear that changing plant genes in unnatural ways could produce weeds that are hard to get rid of.

Some people feel so strongly about genetic engineering that they protest against it.

CASE STUDY: GM Crops in Mexico

Some farmers, especially those who run **organic farms,** choose not to grow genetically modified (GM) crops. They want their crops to be completely natural and without **genes** that have been changed by genetic engineering. In a part of Mexico called Oaxaca, there are thousands of small farms growing just enough food for the families that farm the land. When the corn grown on these farms was tested, some of it was found to be genetically modified.

Scientists think that GM corn from the United States that was meant to be eaten might have been grown in Mexico instead. Genes from the GM corn would then have spread to other corn plants as they **bred** together and made new seeds and new corn plants. Once genes from genetically modified plants are released into nature, they usually cannot be removed entirely from the wild.

This field of GM corn is in Illinois. Some GM plants have already spread their genes to wild plants.

Will Plants Ever Disappear?

Plants are so good at finding places to live that they will probably never disappear altogether. However, some types of plants can die out, or become extinct. We know this has happened already because scientists have found traces of plants that grew long ago, but are not found anywhere today. If one type of plant disappears, it can affect a whole **ecosystem,** which is a community of plants and animals and the place where they live. Some tiny creatures that live on the plants may die out. Larger creatures that eat the plants may not have enough to eat.

Scientists believe there are many plant **species** that have not been discovered yet, especially in places like the Amazon **rain forest.** Some of them may contain substances of great value to medicine. These plants may be disappearing when sections of rain forest are cleared or cut down.

Fossils show us that many species of plants have died out in the past.

Why Are Forests at Risk?

Trees are cut down in large numbers to clear land for farming and to supply wood for construction, furniture, and paper. When old, slow-growing trees are cut down, it takes a long time for new trees to grow. Sometimes the forest never recovers. Rain forests are being cut down at an alarming rate. An area almost as large as Colorado is cut down every year. Replanting projects try to replace these trees. However, new forests are often planted using only one or two tree **species** instead of the great variety of trees and other plants that grew in the original forest.

Rain forests make up some of the world's oldest **ecosystems.** This one is in Australia.

Did you know?

Most of New Zealand's **native** forests were cut down to clear the land for farming. Today, the remaining native forests are protected and pine forests are grown to supply the timber industry.

How can we take care of trees and plants?

There are many ways to take care of plants. One way is to stay on paths in the countryside instead of walking over wild plants. Reducing **pollution** would cause less **acid** rain and reduce the damage it causes to plants. Strict control of land clearance and logging, which is cutting down trees, would help to protect forests. We can also make sure that the wood products we buy come from sustainable forests. These are forests that provide a continuing supply of timber.

We can replant trees and other plants that are at risk. Most countries now have national parks where **native** trees and other plants are protected.

Why do some plants have to be controlled?

Plants are sometimes brought into a country where they do not normally grow. They sometimes grow so quickly that they smother other plants. For example, about 300 native plants in the United States are now threatened by plants introduced from other countries.

Giant hogweed is a problem plant in Great Britain.

CASE STUDY: The Millennium Seed Bank Project

The Millennium Seed Bank Project is an international effort to prevent 24,000 plant **species** around the world from dying out and disappearing forever. The seeds of these plants are being collected and stored. If the plants die out, the seeds can be grown and the plants can be put back in nature.

Seeds are sent to the project's headquarters in the United Kingdom, where they are cleaned and checked to make sure they are healthy. They are then dried and frozen. Every ten years, some of the seeds are taken out of storage and grown to check that they can still produce plants.

Seeds of plants at risk of disappearing from nature are collected and stored in case they are needed in the future.

Glossary

acid chemical compound that contains hydrogen and usually dissolves in water. Weak acids have a sour taste. Strong acids eat into other materials. Oranges and lemons contain citric acid. Your stomach contains acid to help digest food.

breed to produce or increase a plant or animal by sexual reproduction. A person who breeds is called a breeder.

cereal any type of grass that produces seeds or grains that can be eaten

drought long period of time when there is no rain, leaving the ground too dry to grow plants

ecosystem all the plants and animals that live in a place, together with their surroundings

extract take something out of something else, such as coal out of a field

fertile producing much vegetation or crops

fossil remains of a plant or animal that lived many years ago

fossil fuel fuel such as coal, oil, or natural gas made from the remains of living things

fungus type of living thing including mushrooms and toadstools that feeds on plants or animals. More than one fungus is called fungi.

gene unit of a substance called DNA that contain the instructions for the development of a plant or animal

native something that has lived in a place for a very long time and has not been brought in from somewhere else

Native American person who first lived in North and South America before European settlers arrived

organic farm farm where plants or animals are raised or grown using natural methods

organism living plant or animal

photosynthesis process by which plants form food by using carbon dioxide and water in the presence of light

pollution harmful or poisonous substances in nature, usually produced by the activities of humans

process to change a material by using a series of actions or treatments

rain forest woodland with a high yearly rainfall and that is often found in hot parts of the world

sap watery substance found inside plants that carries food and nutrients

shoot part of a plant that grows above ground

species group of related plants or animals made up of individuals that are able to produce young

virus tiny organism or protein molecule that causes disease

More Books to Read

Chambers, Catherine. *Drought.* Chicago: Heinemann Library, 2000.

Galko, Francine. *Classifying Flowering Plants.* Chicago: Heinemann Library, 2004.

Galko, Francine. *Classifying Nonflowering Plants.* Chicago: Heinemann Library, 2004.

Spilsbury, Richard, and Louise Spilsbury. *Plant Growth.* Chicago: Heinemann Library, 2003.

Spilsbury, Richard, and Louise Spilsbury. *Plant Habitats.* Chicago: Heinemann Library, 2003.

Spilsbury, Richard, and Louise Spilsbury. *Plant Parts.* Chicago: Heinemann Library, 2003.

Spilsbury, Richard, and Louise Spilsbury. *Plant Products.* Chicago: Heinemann Library, 2003.

Spilsbury, Richard, and Louise Spilsbury. *Plant Reproduction.* Chicago: Heinemann Library, 2003.

Trapp, Clayton. *Polluted Air.* Chicago: Raintree, 2004.

Index